NORTH TYNESIDE LIBRARIES

3 8012 01768 9028

GW01157279

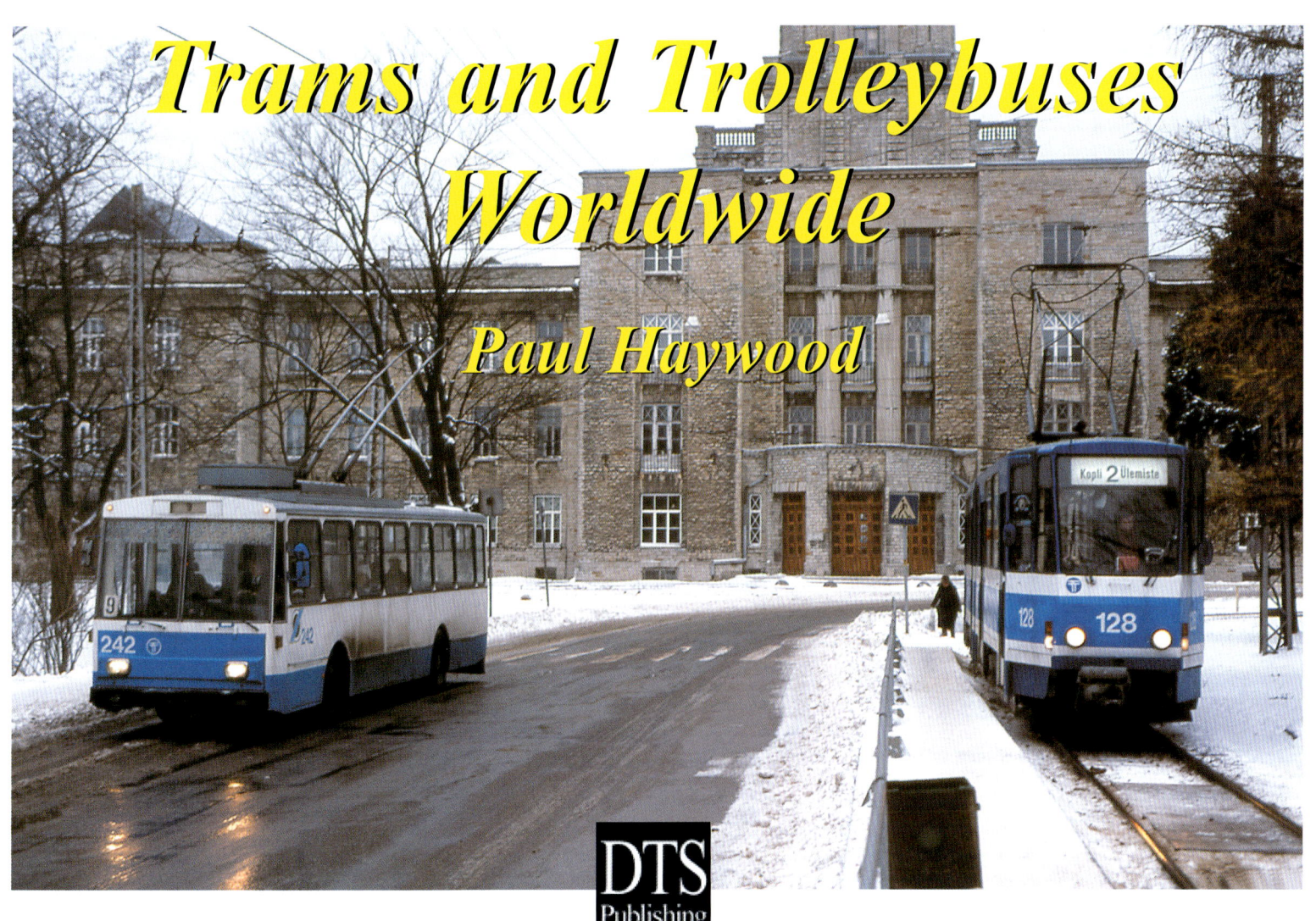

Trams and Trolleybuses Worldwide

Paul Haywood

DTS Publishing

ISBN: 1-900515-41-5   © 2006 Paul Haywood
Published by: DTS Publishing, PO Box 105, Croydon, Surrey

Printed by: Ian Allan Printing Limited

British Library Cataloguing in Publication Data. A catalogue record for this book is available from the British Library.

All rights reserved. Except for normal review purposes, no part of this book may be reproduced or utilised in any form or by any means, electrical or mechanical, including photocopying, recording or by an information storage and retrieval system, without the prior written consent of DTS Publishing

 www.dtspublishing.co.uk

FRONT COVER: San Francisco is justly famous for its cable cars, but also has trams, trolleybuses and buses (MUNI - Municipal Railway) and a metro (BART - Bay Area Rapid Transit). Tram **1214** (built by Boeing Vertol in 1980) meets an unidentified trolleybus (built by Flyer circa 1976), at the junction of Church and 14th Streets in September 1989.

REAR COVER: Budapest tram **2577** meets Schaffhausen trolleybus **207** at the erstwhile Transperience museum in **Bradford**. This short-lived museum opened in 1995 and was intended to be a major attraction featuring historic transport exhibits with working trams, trolleybuses and buses, but sadly the public did not support it sufficiently and the site closed just months after this view was taken in 1997.

TITLE PAGE: Estonia's only tram and trolleybus systems connect at Kopli terminus in **Tallinn**. Skoda 14Tr trolleybus **242** meets (ex-Gera, Germany) Tatra KT4 tram **128** at the terminal loop on a cold day in November 2001

RIGHT: Rome embarked on a tram modernisation programme in the early 1990s, gradually replacing most of their pre-war and PCC bogie cars. Semi-low floor car **9012**, built by Socimi in 1991, negotiates one of the imposing Roman arches at Porta Maggiore in September 2005.

OPPOSITE PAGE: The beautifully tiled Igreja do Carmo church in **Porto** (Portugal) acts as a stunning backdrop for tram **201**, a Brill car of 1929, seen in August 1981. After a nervous few years when the Porto tram system seemed doomed to closure, some routes and heritage trams have been retained and a new light rail system has been developed.

Introduction

Welcome to "Trams and Trolleybuses Worldwide" - a selection of photographs taken by me over the past thirty-five years in thirty countries. This book is not intended to be a definitive history of trams or trolleybuses, nor is it a representative cross-section of the systems and vehicles operated in those countries illustrated. That would be impossible within the limitations of 112 pages (even if I were qualified to do so).

However, I hope these illustrations rekindle happy memories for those readers familiar with the locations, or help to stimulate interest for those who wish to know more about the systems and vehicles illustrated.

Being born in the early postwar years in Leeds, trams dominated my childhood memories. Following their demise in 1959, I made numerous trips to Sheffield to sample their remaining route before they became history a year later. Luckily, a holiday took me through Glasgow the following year, so I was able to ride to the exotic-sounding Auchenshuggle before that system closed in 1962. So, apart from Blackpool, thus ended tram operation in Britain.

However my experience of trams overseas came earlier, thanks to a school trip to Holland in 1959. I distinctly remember telling my somewhat bemused teacher how foolish it was for Leeds to be abandoning its trams in the November of that year, as we rode on one of the new Beijnes 6-axle articulated trams through the streets of Amsterdam. Even then, I *knew* it was folly for us to turn our back on modern trams. The reaction of my teacher was also my first experience of an attitude which unfortunately still prevails in Britain to this day, i.e. that the "progress" of bus replacement was surely more sensible than adopting those strange "foreign" modern trams.

During the 1960s, the trolleybus was a partial substitute for the lack of trams. Yorkshire was then still blessed with systems in Bradford, Huddersfield, Doncaster, Mexborough, Rotherham, Hull and Tees-side. Little did I realise that by the end of that decade most of these would also be gone. The end came when Bradford finally gave up the struggle in 1972.

Back in the 1930s the Leyland bus makers' publicity machine urged tram operators to "Bury your tram with a Titan". Now, in the 1960s, it seemed that the slogan was "Cremate your trolleybus with an Atlantean".

It seems incredible now that such small undertakings as Doncaster or Tees-side should operate a mode of transport now considered only suitable for large cities with good environmental credentials (and finance).

And so ended the trolleybus era in Britain. Now, the only way I could maintain my interest in trams and trolleybuses (outside Blackpool and museums) was to venture overseas. Consequently, in the early 1970s I started to travel abroad, camera in hand, to recapture the joys of tram and trolleybus travel.

Tramways in Europe in the 1970s were in transition. Although there were still many systems with a distinctly pre-war atmosphere, using 4-wheel cars pulling trailers on street tracks, many were upgrading to modern standards by segregating their lines away from other traffic, operated with new, high-capacity articulated trams.

In Western Europe, the dominant technology was German, and Duewag/Siemens were market leaders in modern tram development. Ironically, communist Eastern Europe (the old Comecon countries) used the pioneering PCC technology developed in the USA, through the Czech tram maker CKD-Tatra, with Polish and USSR-made variants. The Belgian maker La Brugeoise (later becoming BN) held a licence to build PCCs that helped a number of Western European systems stay alive beyond the 1960s. In the 1980s, French and Italian manufacturers came on the scene, greatly assisting the renaissance of the tram in their own countries, where the presence of a domestic industry helped to lubricate the political planning and funding process.

In North America, most of the remaining systems used the highly successful but aging PCC streetcars, with technology dating from the 1930s. In the 1970s, an attempt was made by US aircraft builder Boeing-Vertol to build a new "standard" American tram (only bought by Boston and San Francisco), but their poor performance opened the door to imported technology. German, Japanese, Italian, Canadian, French, Spanish and now Czech tram makers have all taken a share of the revitalised US tram market, mostly through local manufacturing deals.

In the 1970s it became clear that the words "tram" or "streetcar" were no longer an accurate reflection of the way the mode had developed, so the US term "light rail" came into common usage.

Light rail had developed into a mode that bridged the gap between highway-based buses and trolleybuses, handling low to medium passenger flows, and the totally segregated heavy metro suitable for high passenger flows. The fundamental essence of light rail was its *flexibility* to fit into any townscape, topography or traffic condition.

Ironically, it was the perceived *inflexibility* of the tram that spelt their doom in the early post-war years in many parts of the world, but that was before the benefits of segregation and modern technology became evident.

Thankfully, the light rail message has now been recognised in most parts of the world, and modern trams are returning to, or appearing for the first time, in towns and cities where even the most optimistic enthusiast could never have dreamt to see them. Virtually every month, a new system or route opens somewhere in the world, and existing systems go from strength to strength.

In those dark days of the 1970s, I never thought I would see trams back in Britain's streets, where they are now helping to regenerate the inner cities of Sheffield, Croydon, Nottingham and others. This welcome, but politically half-hearted, renaissance was long awaited, but I believe we are still the poor relation in the transport world compared with most other developed countries. Even as I write, news of the introduction of the Wright "Streetcar" bus is getting media hype simply because it is a bus with a tram-style articulated body, being promoted as a cheap alternative to the tram. Frustratingly, this was soon followed by the news that a government claiming to support the Kyoto Protocol had scrapped the Leeds, South Hampshire and Liverpool tram projects.

The trolleybus in Britain has been given even shorter shrift. After a brief trial with a French trolleybus in Blackpool in 1983 (see page 43), and a fruitless experiment in Doncaster in 1985, the trolleybus in Britain has been almost forgotten, although the latest funding crisis for new tramways may well divert attention towards this next-best environmental option.

Elsewhere, the trolleybus has had mixed fortunes. The former Comecon countries and China have by far the largest number of systems, often acting as suburban feeders to tram and metro lines. (Significantly, Shanghai is now considered to have the world's longest continuously operating system, having started in 1914). However, deregulation, the increasing availability of oil and the comparative first-cost cheapness of motor buses will undoubtedly lead to some trolleybus (and tram) closures. Before clean air became a fashionable cause, the trolleybus was, like trams in earlier days, considered by many to be an inflexible anachronism, having no advantage over the motor bus other than being powered by locally generated electricity. Even the trolleybus's once cherished hill-climbing abilities were soon eroded following improved tram and motor bus performance. However, the mode is now regaining ground in many places where environmental factors prevail, with dual-powered versions allowing greater route flexibility.

The proportion of tram to trolleybus illustrations in this book is about four to one in favour of the tram. I apologise to trolleybus enthusiasts. This does not imply any lack of interest in the trolleybus, simply that there were more trams to see and photograph in the places I visited. Similarly, I have been able to draw on far fewer sources of information regarding non-British trolleybuses than trams, but hopefully this will change in years to come.

As you will appreciate, many things have changed over the thirty-odd years covered by this book. Consequently, my captions tend to reflect the situation that existed at the time of the photograph, making occasional reference to what may have happened in the ensuing years. Regarding the trams, I have tried to select a variety of vehicle types, old and new, and a cross-section of operating environments, to illustrate the varied nature of tram and light rail infrastructure.

I hope readers will forgive my slightly excessive coverage of UK and US systems, but I feel that all the views have a story to tell.

Finally, my photographs are presented alphabetically by country. So, on page six we start in Australia and on page 112 we finish in the USA.

I hope you enjoy the ride.

Acknowledgements:

Since 1970, I have been a loyal member of the Light Rail Transit Association, and their monthly journal "Tramways and Urban Transit" is essential reading for worldwide news of tram and light rail developments. Their website ***www.lrta.org*** is a treasure trove of information for further research.

I am indebted to Mike Davis for his invaluable help in the skillful preparation of this book.

Paul Haywood
North Yorkshire
August 2006

AUSTRALIA

Melbourne & Metropolitan Tramways Board's once extensive fleet of "drop centre saloon" trams had twin trolley poles which eliminated the need to turn them at journey's end, but still required hooking and unhooking depending on direction of travel. SW6 class tram **944** was built in 1949 in the MMTB workshop. Storm clouds were gathering at Northcote terminus in February 1993 as the conductoress quickly ties off the trolley pole in preparation to return on the short route 9 to Thornbury.

AUSTRALIA
Like many cities in the 1960s and 70s, **Melbourne** was torn between the easy option of scrapping its trams or thinking long-term and modernising. Thankfully, they made the positive decision to modernise. This early example of a route extension was to East Burwood, opening in 1979 (and recently extended further). MMTB **122** is a Z3 class built in 1980 by Commonwealth Engineering (with Duewag trucks) and represented the new face of Melbourne trams when seen on Burwood Highway in October 1982.

AUSTRALIA
There aren't many places in the world where you can still ride on something resembling a pre-war American interurban. These magnificent machines still grace this street in Glenelg on the **Adelaide** tramway where **374**, built in 1928 by Pengelly, headed a two-car set back to the city in February 1993. This line is now to be upgraded with new rolling stock and an extension of the line to Adelaide's railway station.

AUSTRIA
Vienna has the distinction of being second only to St Petersburg in the world rankings of tram system size. However, unlike St Petersburg, Vienna did not neglect its tram system in favour of metro expenditure. SGP built **4030** (with matching trailer by Rotax) descends to Rodaun in October 1988 over track built to light rail standards.

AUSTRIA

LEFT: Linz has a city tramway and a separate "mountain" line called the *Postlingbergbahn*. It is one of the world's steepest tramways using adhesion (without the aid of rack and pinion) with gradients up to 10.6%. Of added interest is their use of Roman numeral fleet numbers and the trolley pole (unusual in Germanic countries). Car **XIV** is a 1954 rebuild of equipment originally built for the opening of the line in 1898, and is seen arriving at the attractive top station after its ascent through the woods from Urfahr in October 1988. Recent proposals to modernise this much-loved line have met with strong opposition.

ABOVE: Innsbruck has two rural light railways, now incorporated into the city's tramway system. The *Stubaitalbahn* is an attractive 18km route to Fulpmes (via the delightfully named villages of Natters and Mutters). IVB **82** is an ex-Hagen tram built by Duewag in 1960, (with a centre section from Bielefeld), seen here entering the stub terminal at Fulpmes in April 2004.

BELGIUM
Gent 3-axle cars **345** and **321** were built by Ateliers de Nivelles in 1935, and were typical of the fleet until the early 1970s, when they were gradually replaced by double-ended PCCs. About this time, following the elimination of an at-grade railway crossing, an opportunity was taken to relocate the former kerbside line at Arsenaal onto centre reservation. Sadly, the connecting shuttle route onward to Melle, still using the kerbside alignment, was earmarked for closure, so this 1972 scene soon became history.

BELGIUM
Gent's transport network, now part of the (Flemish) *de Lijn* organisation, includes a single trolleybus route, opened in 1987. In May 1996 we see Van Hool AG280T **20** at Mariakerke.

BELGIUM
In an attempt to create a regional *semi-metro* network in the **Hainault** region (using light rail technology), the SNCV *(Vicinal)* line from Charleroi to La Louviere via Anderlues and Binche was upgraded with new trams, relaied track and catenary overhead in the 1980s. Sadly, soon after this photograph of BN-built car **6126** was taken near Peronnes in August 1992, the section beyond Anderlues was abandoned, ending what many considered to be the last example of a typical *Vicinal* route through the countryside.

BELGIUM
Unfortunately, because of gold-plating, ill-considered routing and confused politics, many of the *Vicinal's* plans to create a *semi-metro* around **Charleroi** came to nothing, or were built but immediately mothballed. This 1977 view shows SM- class tram **9143** (rebuilt for *semi-metro* high-platform loading), coming off Charleroi's soon to be replaced street alignment near Eden onto one of the *semi-metro* sections actually built (and still in use).

BELGIUM
Brussels embarked on extensive tram subway building activity in the 1970s when many central area routes were closed and relocated below ground (with stations) to be operated as a *pre-metro* for later conversion to full metro. In 1975 we see STIB trams **5008** and **5015** negotiating temporary tracks at Brussels Nord railway station with the earthworks of the tram subway well in evidence. Dyle and Bacalan built these imposing bogie cars to provide high-capacity transport for the Brussels World Fair in 1935.

BELGIUM
Like Brussels, **Antwerp** also relocated some central street lines into subways in the 1970s. Antwerp **2077** is a PCC built by La Brugeoise in 1965 to a design similar to those supplied to Brussels, Marseille, St Etienne and Gent. In May 1975 we see the difficulties of keeping the track points clear during subway construction outside Centraal station.

BELGIUM
The coastal line of the nationwide *Vicinal* (now part of the Flemish *de Lijn* group) once handled a huge amount of holiday traffic in the summer months, with Belgian and German visitors arriving by train, and Britons arriving by ferry. To handle the significant amount of registered baggage, the SNCV/NMVB ordered three powered baggage cars from Godarville in 1932. This September 1971 scene at **Ostend** shows **10021** being used as a shunter for the then still common trailer operations.

CANADA
The **Toronto** Transit Commission was an enthusiastic user of the American Presidents' Conference Committee (PCC) streetcar, developed in the 1930s. For many years, until the late 1970s, their entire service fleet consisted of a mix of newly purchased Canadian-built or second-hand US-built versions. **4310** was bought new in 1946 from the Canadian Car and Foundry Company (to a St Louis Car Co. design), and was seen in mid-wash at Russell depot in 1973. To the left is **4659** - a former Cleveland car built by the Pullman Car Company in 1946.

CANADA

For such a large system, **Toronto** only has a limited amount of reserved track. One of these sections is on Queensway, where ex Cleveland PCC **4698**, built in 1946 by the St Louis Car Co., approaches Humber terminus in October 1973. Note the standee windows, so typical of the North American post-war period.

CANADA
Toronto also had trolleybuses until 1993. In an attempt to modernise and extend the life of their fleet in the early 1970s, components were taken from second generation trolleybuses dating from the 1940s and given new bodies by Western Flyer of Winnipeg. TTC **9334** is seen at the Eglinton subway interchange in October 1973. Note the water-filled rubber bumpers.

CANADA
Following the success of nearby Edmonton's pioneering North American light rail line, **Calgary** wisely opted to use the same tried and tested Duewag U2 light rail vehicles for its own system when it opened in 1981. A mix of street, segregated and ex-railroad alignments were used to create their network in this sprawling conurbation. This single unit was acting as a Training Car when seen climbing out of the city at Franklin in 1989, with downtown Calgary and the distant Rocky Mountains forming a dramatic backdrop.

CANADA
Canada once had no fewer than 14 trolleybus systems, now reduced to two (Edmonton and Vancouver). **Calgary 449**, a Canadian Car & Foundry/Brill type T-44 of 1948, was seen at Thorncliffe in October 1973, less than two years before system closure.

CHINA

There are now few tramway systems left in China, but the north-eastern city of **Dalian** still has a three-route network, and plans light rail expansion. **2009**, built in 1956 in their own workshops and rebuilt in the 1990s, heads out of the city on route 201 in September 2000

CHINA
Dalian also operates trolleybuses, but at the time of this scene only route 101 was in operation. **3948** (possibly built by Shenfeng) attempts to enter the congested main railway station terminal loop in September 2000.

CHINA – Hong Kong Special Adminstrative Region
Hong Kong's famous tramway runs parallel to the island's north coast. Tram **115**, built by Taikoo Dockyard in 1955, passes typical harbour cargo lighters near Kennedy Town in April 1989, with the hills of mainland Kowloon in the background. The then virtually fleet-wide policy of all-over advertising has since been reversed in favour of a reintroduction of the traditional Brunswick green livery with smaller adverts. Subsequent landfill into the bay at this point has changed this view considerably.

Hong Kong's trams are based on a design dating from 1949. In 1986, a number of cars were rebuilt to a less austere style and this soon led to a decision to replace most of the fleet with new trams based on these modifications. Between 1987 and 1992, two local manufacturers were chosen to do this work (Full Arts and Leeway) using trucks and electrical equipment refurbished from the scrapped cars. Cars **48** and **17** (of 1987 and 1989 respectively) are examples of these new cars, now sporting the more traditional livery, at the Happy Valley terminus in 2000.

CHINA – Hong Kong SAR
The Kowloon-Canton Railway Corporation operates the **Tuen Mun** Light Railway, The world's busiest second-generation tramway. The network serves the new towns of Tuen Mun and Tin Shui Wai, and the expanded market town of Yuen Long, all situated in Hong Kong's North-West New Territories. The photograph shows car **1001** approaching Siu Hei stop in southern Tuen Mun during February 1995, and also illustrates the typical multi-storey housing blocks that nourish public transport in Hong Kong. Car 1001 was the last-built of the original batch of Comeng vehicles (1001-1070), because the original No.1001 was damaged during stress testing by the manufacturer in 1987.

CHINA – Hong Kong SAR
"Park and Ride" usually refers to the parking of cars. Hung Shui Kiu station on the **Tuen Mun** system seems to have become an unofficial "Bike and Ride" location as Comeng-built **1035** departs in February 1991.

CZECH REPUBLIC
The Czech company CKD-Tatra built a huge number of PCC-inspired trams, and this T3 of 1983 was typical of those supplied to the former Comecon countries of Czechoslovakia, East Germany, Romania, USSR and Yugoslavia. This coupled pair, headed by ex-Most car **35**, is seen in **Liberec** in February 2004. Note the dual-gauge tracks of this former metre-gauge system that was gradually converting to standard gauge (1485mm).

CZECH REPUBLIC
Prague was the location of the CKD-Tatra tram factory, so the city naturally used its products. This 1962 T3 (originally numbered 6164) was converted to a Driver Training car in 1985 and renumbered **5505**, and was seen on driver training duty demonstrating how easily it could cope with the steep hill up to Hercovka in June 1997.

CZECH REPUBLIC
Prague has long recognised the tourist potential of its trams. Their large collection of preserved cars ensures a varied selection of vehicles available for tourist route 91. Tram **2210**, built in 1930 by Praha-Smichov, is seen leaving Stresovice depot to take up summer tour duty in June 1997.

ESTONIA
Tallinn has the only tramway in the now proudly independent country of Estonia, and is unusual in having a gauge of 1067mm. When the country was part of the former USSR, they were left with little choice but to buy Tatra products from Czechoslovakia. Fortunately, these vehicles were extremely rugged, fast and reliable. Coupled Tatra T4s **290** and **302** pass articulated KT4 **69** at Lubja in November 2001.

EGYPT

LEFT: Cairo's ill-maintained tram system has virtually disappeared since this view was taken at Bab el Shara in 1991. Typical of the state of their fleet was twin-car **4156**, built by Kinki Sharyo in 1982. (The driver of the horse-drawn tanker had just accepted the inevitable and realised that the on-coming tram was not able to change direction, after a stand-off lasting some five minutes.)

ABOVE: In spite of a considerable number of new lines built and rolling stock purchased in the 1970s and 80s, most of **Cairo's** tramway is now just a memory. Kinki Sharyo-built **4020** was seen at Shubrah terminus in 1991, where the adjacent roadway seemed to have been requisitioned by the Permanent Way department. Note the *'Arret'* tram stop - an echo of a time when the Cairo tramway was once Belgian-owned.

EGYPT
The **Heliopolis** tramway still links that city with Cairo and is mostly segregated from other traffic. 3-car set **809**, built in 1976 by Kinki Sharyo, arrives at Cairo's Ramses Square with an appropriate statue of King Ramses II dominating the scene. At the time of this view in September 1991, the set continued to Corniche, but the route has since been cut back to this location.

FINLAND
Helsinki has a medium-sized tramway and a small metro system. This solid-looking vehicle represented one of the last traditional bogie trams purchased prior to the introduction of articulated cars in the 1970s. **19** was built in 1959 by Karia/Valmet and was seen at Kauppatori in July 1988.

FRANCE

ABOVE: Marseille's single tram line is, at the time of writing, closed for rebuilding. This pair of unidentified La Brugeoise-built PCCs of 1968 (and refurbished by BN in the 1980s), exits the historic city centre subway in September 1994.

RIGHT: Paris has enthusiastically reintroduced the tram to its outer suburbs to act as metro feeders. The *Val de Seine* route T2 in the west of the city links La Defense and Issy using a former suburban railway alignment. This line (with earlier line T1) is spearheading a significant number of new tram routes now being planned and built in the city. RAPT **209** was built in 1996 by Alsthom to a Grenoble design with one of the earliest successful low-floor layouts, seen at Les Coteaux station, with its tastefully adapted railway-era architecture, in May 1998.

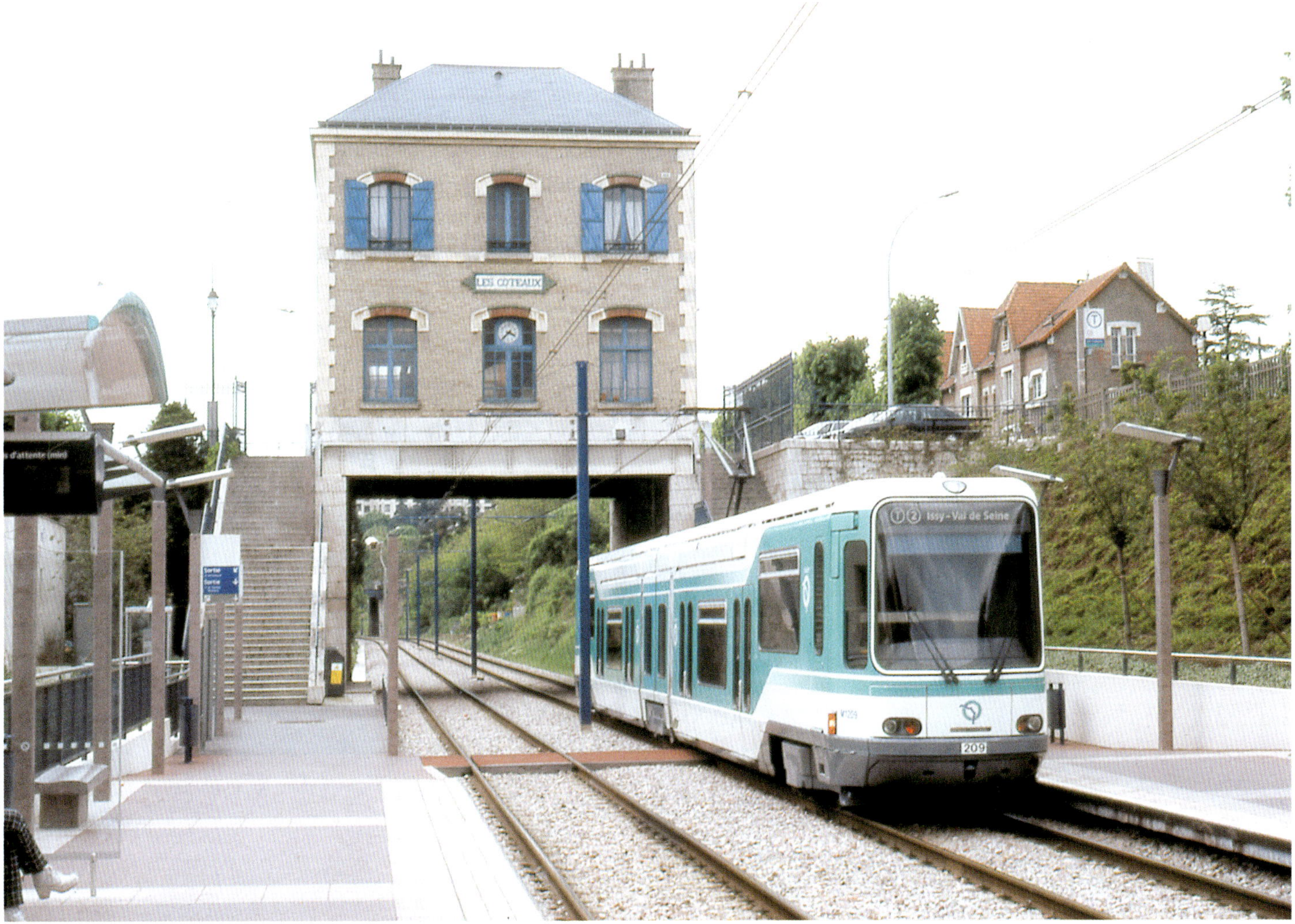
LES COTEAUX
Issy - Val de Seine
209

FRANCE
The interurban tramway linking **Lille** with Roubaix and Tourcoing has been gradually upgraded to light rail standards. Back in March 1973, tram **516**, built by Brissoneau & Lotz in 1949, represented a batch of cars that would continue to be the mainstay of the fleet until the 1980s. This section of route into the city has since been replaced with two separate tunnels (one replacing the other).

FRANCE
An excellent example of one of the many second generation tram systems opened or being built in France is **Montpellier** which commenced service in 2000. This scene shows a new connecting junction being laid for the city's second route, due to open in late 2006, as Alsthom Citadis 8-axle tram **2016** sweeps round the corner outside Gare Saint-Roch in March 2006.

FRANCE
St Etienne has one of the five remaining trolleybus systems operating in France - the others being the hybrid systems of Caen (TVR "tram on tyres") and Nancy (guided trolleybus) and Limoges and Lyon ("traditional" systems). **406** is a Berliet ER100 and is seen outside the city's Châteaucreux railway station in March 1983.

FRANCE

In November 1983, to form part of a transport conference held in Blackpool, the French city of **Nancy** sent one of their new Renault PER180 dual-powered trolleybuses (**610**) for demonstration to delegates. To showcase the vehicle's flexibility, delegates were transported from the conference venue by diesel power, then converting to trolleybus mode using a short, specially installed, twin-wire section near Blundell Street tram depot. Needless to say, nothing came of the demonstration. These vehicles have since been replaced in Nancy by a hybrid, guided trolleybus with a tram-like body using a single (sunken) guide rail for part of its route.

GERMANY
A Trabant car on a cobbled street, with a 4-wheel tram running on a single, kerbside track. Representing a bygone age in October 1990, a few days after the official reunification of East and West Germany, was **Woltersdorf 31**, built by Gotha in 1960 for the Schwerin tramway.

GERMANY
The smart, clean lines of the Duewag *Stadtbahn M* metre-gauge tram never seems to date, even though the design is now thirty years old. The location is the Sterkrade terminus of the one-route **Oberhausen** system which reopened in 1996. However **274**, built in 1976, belongs to the adjacent city of Mulheim who operate the route jointly, and was seen preparing to return to its home territory in May 2006.

GERMANY
Nearly all cities and major towns in Germany have a tram or light rail system, but only a handful have trolleybuses. Perhaps the best known system is in **Solingen**, which has a five-route network. Route 683 connects the famous *Schwebebahn* (suspended monorail) at Wuppertal Vohwinkel with the delightful village of Burg, where the restricted terminal location requires the use of this turntable. Because of its diameter, the route is restricted to the use of non-articulated vehicles like **32**, built in 1986 by MAN/Graf/Kiepe, seen in May 2006.

GERMANY

When **Stuttgart** decided to upgrade their metre-gauge tram network, they opted create a standard-gauge *Stadtbahn* (light rail) system using tunnels and segregated alignments, eliminating much traditional street running. This scene shows metre gauge tram **423** (built in 1960 by Esslingen) passing the newly emerging *Stadtbahn* line and station at Ruhbank in June 1996.

GERMANY
One German system had the enterprise to develop a new and highly potential use for light rail. Given the right conditions, **Karlsruhe** knew that the modern light rail vehicle could operate perfectly safely and economically on existing suburban rail lines, opening up the possibility of through-running between tramway and mainline railway. Dual-voltage Duewag tram **813** of 1993 is seen at Karlsruhe's Hauptbahnhof in June 1996.

GREECE
Athens' trolleybus system has recently been re-equipped with Van Hool and Neoplan vehicles, replacing these Russian-built ZIU-9s, of which **2132** was a typical example in January 1990.

HUNGARY
The river Danube separates the twin cities of **Buda** and **Pest**. On the Pest side of the city in April 1993 is tram **1327**, a Ganz-built car from the late 1960s, passing a moored hydrofoil with Buda's famous Gellert hotel and spa seen on the opposite bank.

HUNGARY
Budapest, like many former Comecon cities, has a trolleybus system. Ganz, the Hungarian electric traction company, teamed up with bus maker Ikarus to create this trolleybus version of the popular 280 series. **240**, of 1989, is seen at Ors Vezer teren in April 1993.

IRISH REPUBLIC

Dublin reintroduced the tram to the city in 2004, after a gap of 55 years. The first of two lines to open was the Green Line running south to Sandyford that makes considerable use of a former railway alignment. Marketed as *Luas* (Gaelic for *speed*) the two lines are currently unconnected, but plans are being made to link them through the city centre. Green Line **4014** is an Alsthom Citadis to an 8-axle configuration, (the Red Line uses a 6-axle version). The excitement and curiosity created by the trams is evident in this first day scene at St Stephen's Green on 30 June 2004, when crush loads were experienced well into the evening.

ITALY
Milan bought trams in large numbers in the 1920s to an American Peter Witt design, and their high capacity and passenger flow layout revolutionised tram travel. Their rugged design and serviceability is evidenced by the fact that some of them have been refurbished and are still in service nearly 80 years later. ATM **1508**, built by Carminati e Toselli in 1928, heads a line-up of sister cars at Viale Sabotino in November 1988. This tram queue was the result of a street closure caused by broken Christmas lights threatening to touch the tramway overhead, requiring the attention of the Fire Department.

ITALY

Milan's imposing Fiat 2418/Viberti trolleybus **624** of 1959 emerges from beneath Stazione Centrale in August 1984. Note the unusual 3+1-axle configuration, and the right-hand driving position.

ITALY
Rome is justly famous for its ancient ruins. In August 1984, the Coliseum was undergoing extensive renovation when car **2093** (built by Carminati e Toselli in 1932) passed this much-photographed location.

JAPAN
The Nikkoku Kogyo company built **6080** in 1949 for what was then **Tokyo's** extensive tramway system. Forty years later it was on static display at Oji Park during the cherry blossom season, which is always regarded as a good excuse for family picnics. The Japanese sense of civic responsibility has ensured the tram has not fallen victim to graffiti or vandalism.

JAPAN
Tokyo still retains two separate tram routes in this subway-rich city. The *Arakawa* line runs mainly on private right-of-way but includes two street running sections. Cars **7020** and **7008** (both built in 1955 by Nippon Sharyo/Hitachi, and rebuilt in 1977 by Aruna) are seen at the end of one of the street sections at Oji in April 1989. Note the doors that require high platforms.

JAPAN
In February 1991, these beautifully maintained trams were seen at Sumiyoshi junction on the two route *Hankai* (**Osaka**) system. The Kawasaki Car Company built **165** in 1928, and **171** was built in 1931 by Tanaka Sharyo to a virtually identical design. Japan's earthquake-prone geology requires that power lines are easily accessible, which doesn't help to improve the urban scenery.

MALAYSIA
Kuala Lumpur never had a tram system, but recently embraced the light rail concept by building two separate and totally different segregated third-rail lines, and one monorail line. The *PUTRA* line is automated (London Docklands-style) running mainly on elevated tracks, with a central-area subway. Leading a two-car set, **134** was built by Bombardier to open the line in 1999, and is seen near Pasar Seni in August 2003. This route is an excellent way to see the city, particularly from the driverless front window. Note the world's one-time tallest structure, the Petronas Twin Towers, to the distant right.

MALAYSIA
The other light rail line in **Kuala Lumpur** is the two-route *STAR* line which is a more traditional (driven) system with a mix of elevated and segregated at-grade alignments. Adtranz –built, 4-car set **1118** is seen approaching Masjid Jamek station with the distinctive outlines of the mosque and Bank Negara dominating the scene in 2003.

NETHERLANDS

Amsterdam's Haarlemmermeer railway to Uithoorn has not seen passenger trains since 1950 or freight trains since 1972. Soon after full closure, *De Stichting Electrische Museumtramlijn Amsterdam (EMA)* took over part of the line to run its collection of preserved trams. The original station building is still in use as a café, bookshop, ticket office and waiting room, but the former stub terminal tracks have been replaced with this turning circle. In May 1995, Vienna **2614** built in 1921 by Graz (with matching trailer) departs on the 7km run to Amstelveen.

NETHERLANDS

Houten, near Utrecht, is the location of an admirable transport innovation. The nearby community of Castellum, on the Utrecht to s'Hertogenbosch railway, is developing rapidly, but does not yet justify its own railway station. Instead of allowing the residents to become isolated and forced to use their cars, the decision was taken to lay a third track alongside the mainline for some 2 kilometres to a one platform halt at Castellum and run a single tram as a temporary shuttle to Houten station. Here we see former Hannover Duewag tram **6016** of 1974 meeting the train at Houten in November 2001.

NETHERLANDS
In **Rotterdam** in 1969, route 5 was extended to Schiebroek to serve a small but growing housing estate via a long tram-only flyover spanning roads, a railway, a canal and (soon after) a very busy motorway. Had this new line not been put in early, building it would have been very difficult and vastly more expensive. RET 600-class tram, built in 1969 by Werkspoor to a Duewag design, speeds over the viaduct on a cold March day in 1972.

NETHERLANDS
The smart coastal resort of Scheveningen is served by no fewer than four tram routes from **Den Haag** (The Hague). Since the early 1950s, Den Haag was a keen user of American-designed PCC trams, built under licence in Belgium by La Brugeoise, but unlike most PCC's of European origin, these cars had US-style bodies. In July 1992, at the attractive terminus of route 11 at Scheveningen Strand, is **1330** (with matching powered trailer) built in 1972, representing some of the last traditional-style PCCs to be built.

NETHERLANDS
The *Floriade* flower festival takes place at different locations in The Netherlands every 10 years. In 1992 it was held in **Zoetermeer** and included a specially built tramway to take visitors around the site. The wealth of museum trams available in the country meant that there was no shortage of cars to operate the service. Den Haag (The Hague) **830** and trailer **756**, were built by La Brugeoise in 1929 and gave valuable service during the 6 months of the event.

NETHERLANDS
Arnhem is proud to possess the only trolleybus system in The Netherlands. **151** was built in 1984 by Den Oudsten/Kiepe and was seen at Velp Noord in July 1996.

NEW ZEALAND
For such a small country, New Zealand is blessed with no fewer than four heritage tramways - at Wellington, Auckland and Christchurch (with two). In 1993, the only tramway in Christchurch was at the **Ferrymead** Heritage Museum where Christchurch Boon car **152** of 1910 gave rides from the entrance to the museum's re-created township. Today, this tram is again running in the streets of Christchurch thanks to the building of a 2.5km city centre loop line using preserved trams.

NEW ZEALAND

Wellington Tramway Museum has a delightful line in Queen Elizabeth Park, situated some 45 kms north of the city. The remoteness of the location is well illustrated by this view of Wellington **159** (built in 1925 by Wellington Tramways to a Boon design) as it travels back from the beach to the depot and park entrance in February 1993.

NEW ZEALAND
Wellington now possesses the country's only trolleybus system. In February 1993, **232** waits for **216** to exit the famous single-lane, signal-controlled, Hataitai trolleybus tunnel (originally built for trams in 1907). Both vehicles were built by Volvo/BBC with Hawke bodies in 1986.

NORWAY

LEFT: In February 1981, **Oslo** was still regularly operating these smart 'Goldfish' cars built by Skabo/AEG in 1939 where **169** was seen outside the city's Sentral railway station.

ABOVE: These single-ended cars had a distinctive and very '1930s' rear end, designed to give the impression of speed and modernity.

PHILIPPINES
Manila opened this elevated light rail line in 1984 using Belgian (BN) vehicles. This two-car set, headed by **1013**, approaches Vito Cruz station in February 1991. Although the segregated route has obvious benefits for travellers, avoiding severe congestion at peak periods, the ever-present Jeepneys still rule the road.

POLAND
These two **Warsaw** trams represent different phases of development of the Konstal PCC design. On the left is **1368**, a 105N of the 1980s and **402** is a 13N of the 1960s converted to a works car, seen at Gorczeweska terminus in June 1994. (Is the author alone in seeing the tourism potential of an open-backed tram service?)

POLAND
Warsaw was severely damaged during World War Two. The Old Town was totally destroyed but rebuilt to near-original facade appearance (as seen) between 1949 and 1963, when through traffic was diverted by means of this tunnel. This pair of Konstal 105Ns headed by **1358**, exchanged passengers in the time-honoured way in June 1994.

POLAND

Warsaw had a trolleybus system from 1946, using Soviet-made vehicles, which closed in 1973. In 1983 in an attempt to re-use the redundant vehicles that had been stored since closure, plans were put in hand to open new routes, but only one line was actually built. By the early 1990s the rolling stock was in dire need of replacement but lack of funds precluded the purchase of new vehicles. As a temporary measure these second-hand Swiss vehicles were bought in 1992, but this did not prevent the line's closure in 1995. Saurer built **T019** (ex St Gallen 126) of 1957, and trailer **P004** (ex St Gallen 331, built in 1970 by Hess/R&J), were seen at Wilanowska terminus in June 1994, still wearing their Swiss livery.

POLAND
Residents of this **Poznan** suburb have this express tramway at their disposal, where coupled Konstal 105N cars **112** and **111** arrive at Polanka to pick up passengers for a fast ride into town in June 1994.

PORTUGAL

ABOVE: Europe's most westerly tramway, the **Sintra**-Atlantico tourist line from Banzao to the coast at Praia das Macas, has recently been re-opened back to Sintra. However, when this photograph was taken near Pinhal de Nazare in August 1981, the line had a very uncertain future (having been closed from 1974 until 1980). Tram number **1**, built by Brill for the opening of the line in 1904, had just experienced a de-wirement followed by a detached trolley retrieval rope. Retying required this precarious ascent by the conductor, much to the amusement of the good-natured staff, onlookers and passengers (including the author's son and late wife).

RIGHT: Lisbon used to take its trams for granted, but now some of the remaining routes are marketed as tourist attractions. The most famous route on the system is the Graca circle with its steep gradients, tight curves and narrow streets. Tram **745**, built in 1947 by Maley & Taunton/Lisbon Electric Tramways, swings round the interlaced-tracked corner at Alfama in February 1990. Note the traditional hand signalling to prevent conflicting traffic movements, since replaced with automatic light control.

77

RUSSIA

St Petersburg (formerly Leningrad) still has the distinction of having the world's largest tramway. Sadly, the system is shrinking and neglected in favour of metro investment, and is now subject to bus competition and minibus piracy. Poor maintenance of vehicles and track, and increasing road congestion, have led to this once proud system having some of the slowest service speeds anywhere in the world. **0416** is an LM-99 built in 2001 by PTMZ, representing one of the few new trams introduced in recent years, enjoying a short stretch of traffic-free track at Sadovaya in March 2005.

RUSSIA
St Petersburg has a large trolleybus system, which is also in desperate need of investment. Trolleybuses of type ZIU-9 were built in huge numbers and also saw service beyond the former USSR. **2805** is seen at St Isaac's Square in March 2005.

RUSSIA
Trolleybus works vehicles are rare, but **6604** in **St Petersburg** was found taking an "off-the-wire" rest in March 2005. The radiator suggests it is dual-powered to allow access to units stranded by power failures.

SOUTH AFRICA
The nearby trees and poles make turning the trolley pole difficult for the staff of **Kimberley** car **1**. This is a 1985 museum rebuild, based on the remains of original Stephenson tram 3, built in 1904 for the opening of the Kimberley and Alexandersfontein tramway. It is seen at the "Big Hole" Diamond Mine museum terminus in November 1998, preparing to return visitors to Market Square on what is the only remaining working tramway on the continent outside North Africa.

SPAIN
The Mallorcan town of **Soller** is linked to its port and beach resort by this delightful 5km tramway. Car **2** and its matching trailers were built by Carde y Escoriaza for the line's opening in 1913. They are seen heading back to Soller to connect with the scenic railway to Palma in July 2003.

SPAIN

The Tibidabo line in **Barcelona** was, for many years, the only place you could ride on a street tram in mainland Spain. Now, modern trams are back in Barcelona, Bilbao and Valencia and other cities are actively planning or building new systems. This tram connects with a suburban branch of the Catalan Railway at Avinguda Tibidabo to take mainly weekend and holiday traffic up the hill to a funicular for the ascent to the Tibidabo amusement park offering magnificent views over the city. Car **6**, built in 1905 by Estrada, was in excellent condition when the ritual of turning the trolley pole was re-enacted at the lower terminus in April 1992.

SWEDEN

Gothenburg's tramway was, in the 1970s, considered to be a showpiece for modern tram (light rail) operation, where many new road or housing developments included the provision of a fast segregated line. This twin-car set, headed by **809** (built in 1969 by Hagglund), speeds out of town at Anggardsplatsen en-route to Tynnerad in February 1981.

SWEDEN
Gothenburg's premier light rail route is to Angered, which includes a deep tunnel with a station. Hagglund-built tram **612** is seen at Hammarkullen's cavern-like underground station loading for the city in February 1981.

SWITZERLAND
Like Innsbruck, **Basel** also had rural light railways that were linked with the city tram system in the 1980s. The former BTB (*Birsigtalbahn*) has the distinction of being an international route, crossing into France en-route to the Swiss terminus of Rodersdorf. Baselland Transport **231** (built by Schindler in 1979 with a newer low-floor centre section) is seen leaving France (at the appropriately named Napoleonstrasse) en-route to Basel in October 1993. The set would then continue over the former BEB (*Birseckbahn*) to Dornach.

SWITZERLAND
Where else other than Switzerland could you find railway, bus, mountain rack railway and local tram integrating so conveniently? At **Bex's** mainline railway station in June 1971, we see Bex-Villar-Bretaye (BVB) mountain railway car **25** (built in 1945 by SLM) with cargo trailer, a Post bus, and BVB town tram **15**, built in 1948 by SWS/SLM for local service to Bevieux.

SWITZERLAND
The wealth of home-generated hydro-electric power allows Switzerland to use electric traction for much road and rail public transport. **Luzern's** N&W/Hess/Siemens trolleybus **253** and trailer arrives at the Hauptbahnhof interchange in March 2006.

SWITZERLAND

The interurban tramway from **Vevey**, via Montreux and Chillon to Villeneuve, on the shore of Lake Geneva, was replaced by trolleybuses in 1958. VMCV **2**, built in 1957 by ACMV, is seen at Vevey railway station in June 1971. These first generation vehicles have now been replaced by articulated Van Hool AG300Ts, of which **13** *(above)* is an example, seen at Villeneuve terminus in March 2006.

SWITZERLAND
Many first generation tramways in Europe started with horse or steam haulage (often both). Surviving steam trams are rare, but this magnificent example has been lovingly restored in **Bern** for use on weekend tours of the city. Loco **12** was built in 1894 by SLM but trailer **31** is a beautiful reproduction built in 2002 by Gangloff/LWB and were seen on a city-centre publicity trip in March 2006.

UNITED KINGDOM
Bradford operated the first and last trolleybus system in the UK, opening jointly with Leeds in 1911 and closing in 1972. In the final decade of operation, most services were operated by a mix of second-hand and newly purchased vehicles dating from the 1940s, but rebodied by East Lancs in the 1960s to extend their lives. Karrier-built **731** was on borrowed time when seen at Buttershaw in 1971, but was saved for preservation.

UNITED KINGDOM

LEFT: Blackpool stills carries the tramway torch, having given continuous service since it opened in 1885 as a pioneering electric operation using the conduit system. During the tramway's Centenary celebrations in 1985, many preserved trams were brought in from museums to add variety to the occasion. Edinburgh **35** was one of a final batch of trams built in their own workshops, with Peckham trucks, in 1948, and only saw eight years service before the system closed in 1956.

ABOVE: The then most modern tram in Britain, **Blackpool 762**, which was rebuilt in 1982 from a 1935 English Electric "Balloon" tram, made an embarrassing contrast with **Nancy** trolleybus **610** at Blundell Street depot in November 1983 (see page 43).

UNITED KINGDOM
The light rail revival in the UK started in **Newcastle** in 1980, with the opening of the totally segregated Tyne & Wear Metro, using former British Rail alignments and a new city-centre subway. **4063**, built by Duewag/Metro-Cammell, loads at Tynemouth on the first day of operation on 11 August 1980. The realignment of the track and platform and the huge differences of scale between the needs of heavy and light rail could not be more evident.

UNITED KINGDOM

The next light rail system to open in the UK was the totally segregated, third-rail powered, Docklands Light Railway in **London**, using driverless vehicles. Untypical of UK transport planning, the railway was put in early to allow patronage to grow in line with commercial and housing development. These three views show the remarkable changes that took place at Heron Quays - looking north towards Canary Wharf - between the first year of operation in 1987, then 1989 and finally 1991. The vehicles illustrated were all built in 1987 by Linke Hoffman Busch, and have since been sold to Essen, Germany where they now operate as driven trams using an overhead power supply.

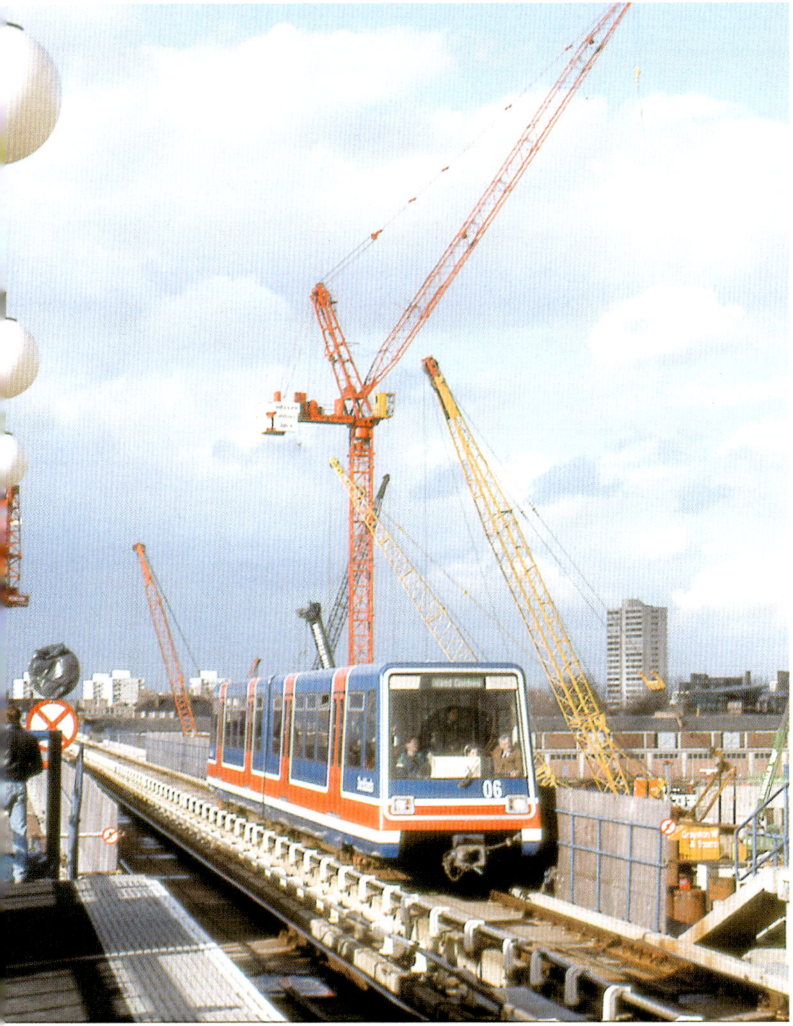
06

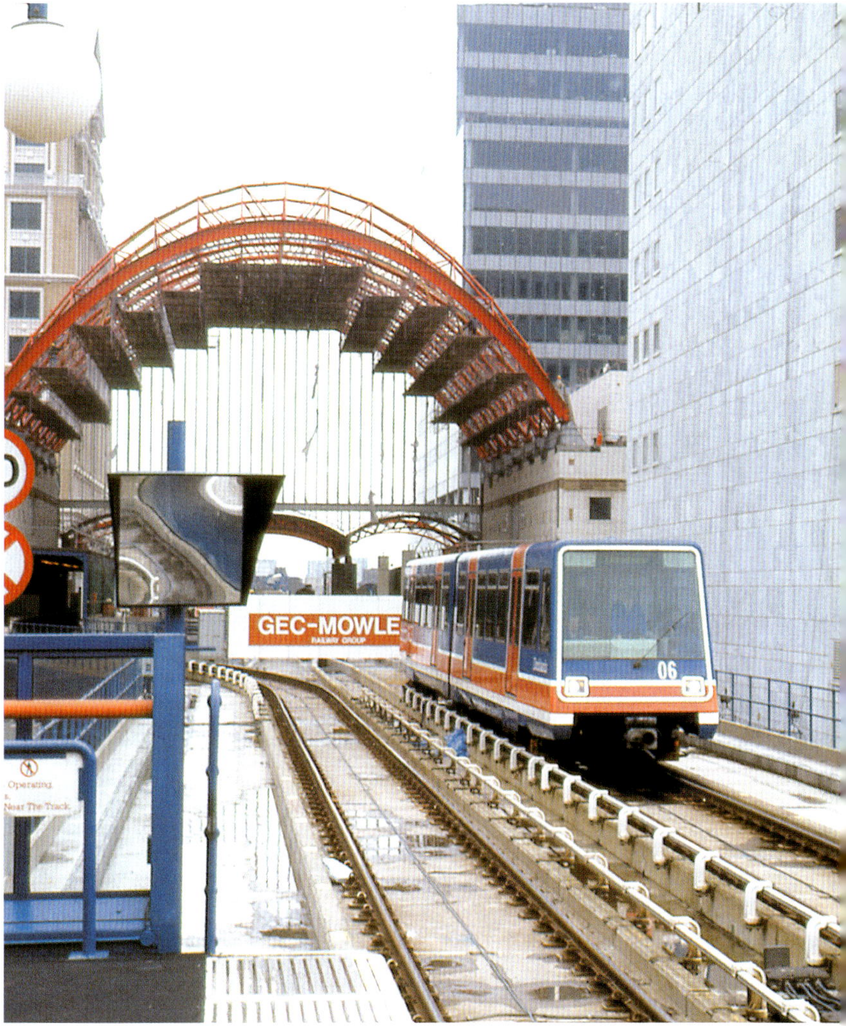
GEC-MOWLE
06

UNITED KINGDOM

"Project Light Rail" was held in March 1987 to demonstrate the mode and its infrastructure to representatives of local and national government, commerce, industry and the media for what was then still a novel concept in the UK. London Docklands **11** (newly-built by German makers Linke Hoffman Busch for third-rail operation) was specially fitted with a pantograph for use over a short stretch of British Rail track at Debdale Park, **Manchester**. Ironically, this vehicle is now back in Germany (Essen) being used as a driven tram (see previous pages).

UNITED KINGDOM

Manchester's Metrolink system opened in 1992, using former railway alignments, taking over the northern Bury and southern Altrincham lines from British Rail, and linked by street running between Victoria and Piccadilly stations. A branch from the Altrincham route to Salford Quays opened in 1999 and onward to Eccles in 2000. Six new trams were delivered to operate the line of which **2006**, built by Ansaldo Firema, was seen near Salford Quays in 2000. Note the shrouded couplers which the Railway Inspectorate deemed to be essential for mixed-traffic operation in Eccles.

UNITED KINGDOM
The South Yorkshire Supertram system opened in **Sheffield** in 1994. Early problems plagued the system such as a fare war with competing buses, the demolition of housing estates along some routes and high levels of vandalism and fare evasion. Thankfully, service speed, superior ride quality and better public relations soon saw ridership increase, and the now Stagecoach-owned Supertram is a well-loved part of the urban scene. Just months after the first section opened from Meadowhall to the City Centre in 1994, tram **14**, built by Duewag, is seen climbing out of the Don Valley at Woodbourn.

UNITED KINGDOM
Next in the UK light rail revival was the Midland Metro, running from **Birmingham** Snow Hill along most of the former Great Western railway route to Wolverhampton. The line opened in 1999 (154 years after the first through trains along this alignment). The obvious railway inheritance of the line can be seen at Bilston Central, where trams **15** and **12**, built in 1998 by Ansaldo, pass at the staggered platforms in 2004.

UNITED KINGDOM

Croydon lost its traditional trams in 1951 but now spearheads the light rail revival in the London area. Their three-route system opened in stages from 2000, using a fascinating mix of street, private-right-of-way and ex-railway alignments. Tramlink **2543,** built by Bombardier, uses a one-way loop within central Croydon and gives George Street a distinctly 'continental' look soon after opening in 2000.

UNITED KINGDOM
In spite of the Labour Government's supposed commitment to light rail development in the UK, **Nottingham** may be the last new system (outside Edinburgh) to open for the forseeable future. The system uses a mix of on-street and former railway alignments. Nottingham Express Transit **203**, built by Bombardier, leaves Cinderhill halt on the short branch to Phoenix Park-and-Ride, on the first day of operation in March 2004 on what is clearly an ex-railway alignment along the former Babbington colliery branch.

UNITED KINGDOM

Crich Tramway Museum represents the benchmark for the preservation and operation of first-generation trams in the UK. Since the site was bought in 1959 (being a former mineral railway built by George Stephenson to serve the nearby quarry), the museum has gradually transformed itself into a recreated Victorian townscape served by a superb collection of preserved (mainly) British tramcars. Glasgow **812** dates from 1900 (and rebuilt to this condition in 1930), passes the depot and Red Lion "pub" in 2001. Note the dominant memorial to the Sherwood Foresters who fell in the First World War.

USA

In 1897 **Boston** made history when it opened the first tram subway in the USA, under Tremont Street. **3661,** built in 1987 by Kinki Sharyo, is seen at Park station during the subway's Centenary year. The open station layout, and the tram's double-sided doors, allow swift loading, unloading and route interchange.

USA

The Electric Railway Presidents' Conference Committee (PCC) streetcar was developed by the Transit Research Corporation in the 1930s to help North American operators combat the increase in motor car competition. By 1950 production had reached nearly 5000 units for the USA and Canada. Most have been scrapped many years ago, but a significant number have been saved for heritage use, particularly for the Market Street line in San Francisco. One other place where they are still in daily use is **Boston**, on the short subway feeder route from Ashmont to Mattapan. **3254** was built in 1945 by the Pullman Car Company and was seen at Cedar Grove in October 1997.

USA
San Diego's El Cajon light rail line uses part of the old San Diego and Cuyamaca Railroad route. This former Mohave Northern switcher locomotive, with a Southern Pacific box car and caboose, were static exhibits at La Mesa Museum to represent the line's heritage, with Duewag U2 car **1026** passing in February 1991.

USA
Philadelphia's former Red Arrow line to Media is a modest remnant of what was once a huge network of interurban lines in the USA. Kawaski-built car **100** of 1980 is seen leaving Media's single-track Main Street to take the private-right-of-way route back to the 69th Street Terminal in October 1997.

USA

The predictable and unvarying path of a tram, with its quiet and emission-free motors is ideally suited to pedestrianisation and city centre regeneration. **Sacramento's** US-built Duewag U2a **105** gently negotiates the downtown loop from K Street Mall into 7th Street in February 1994.

USA
San Francisco's Flyer trolleybus **5193** takes on passengers at Church/30th Street, with track work for the extension of tram route J to Balboa Park in 1989.

USA

The **Los Angeles** Blue Line reuses much of the former Pacific Electric route to Long Beach. **147**, built by Nippon Sharyo in 1990, trails a coupled set into the downtown Flower Street/7th Street subway just days after this section opened in February 1991. After many years of planning delays and financial and political wrangles, the city now has four light rail lines in this once car-infatuated city.

USA

As the sign says – **New Orleans** has the oldest continuously operated street railway in the world, better known as the St. Charles line. Sadly, Hurricane Katrina and the subsequent floods badly affected the trackbed of this route, and the trams of the newer Riverside and Canal Street lines that became water-logged. In happier times we see St Charles line **915**, a magnificent Perley Thomas car of 1923, approaching the Carrollton stop in October 2000.

USA

Portland has rejuvenated its downtown area, thanks in no small part to the introduction of light rail. **109** was built by Bombardier (to a BN design) in 1986 and is seen on the Pioneer Square loop in September 1989. Unfortunately, low-floor technology came too late to prevent the need for this expensive and time-consuming disabled access facility.

USA

The small system of **Missoula**, Montana made streetcar history in 1910 when it opened with system-wide one-person operation, using large bogie cars. (Earlier systems only had partial one-person operation, often only using smaller cars). The company's only interurban car (**50**) – for the Bonner line – was built by Brill/American Car Company in 1912. When the system closed in 1932, most of the fleet was bought by a local camp site for use as holiday accommodation. This fortunate situation allowed car 50 to escape the torch, and the windswept but remarkably intact vehicle was seen at Fort Missoula heritage park in 1982. The tram is now the subject of extensive restoration back to its former glory.